The Banana Lover's Cookbook

by Carol Lindquist

St. Martin's Press

New York

A Nantucket Press Book

Design director: Jon Aron
Editor: Leslie Linsley
Associate: Lisa Sherburne
Illustrations: Jon Aron

THE BANANA LOVER'S COOKBOOK.
Copyright © 1993 by Carol Lindquist. All rights reserved.
Printed in the United States of America. No part of this
book may be used or reproduced in any manner
whatsoever without written permission except in the case
of brief quotations embodied in critical articles and
reviews. For information, address St. Martin's Press, 175
Fifth Avenue, New York, N.Y. 10010.

Library of Congress Cataloging-in-Publication Data

Lindquist, Carol.
        The banana lover's cookbook/Carol Lindquist.
              p.   cm.
        ISBN 0-312-08702-0 (pbk.)
        1. Cookery (Bananas) I. Title
TX813.B3L55 1993
641.6′4772—dc20   92-41399
                                          CIP

10 9 8 7 6 5 4 3 2

For my husband Karl, man of my heart. Together we grew the bunch of bananas that launched this book. And to our dog Yellah, who acquired a taste for bananas in the process.

I also wish to thank the following people and associations:
United Fresh Fruit and Vegetable Association in Alexandria, Virginia, for a wealth of information from the *Encyclopedia of Produce*, by R. A. Seelig and updated by Michelle C. Bing; Laura A. Kinkle, for her kind help; Chiquita Brands International, Incorporated; Dole Food Company; Gregg Thompson at Banana Supply Company, Incorporated, in Miami, Florida; International Banana Association (IBA) in Washington, D.C.; Polly Peck International/Del Monte Tropical Fruit Company in Coral Gables, Florida; and finally, Leslie Linsley and Jon Aron, with very special thanks for turning the banana idea into the banana reality, and for enduring my experimental four-course banana dinners.

Contents

Soups, Salads & Sandwiches

Hearty & Savory Dishes

Cakes & Cookies

Puddings & Pies

Other Desserts, Frozen to Flambéed

Kids' Treats

Foreword

According to Hindu legend, Adam and Eve succumbed, not to an apple, but to a banana. There is no denying the allure of a perfectly ripened, sweet banana. In fact, so powerful is its attraction that today the banana ranks as the favorite fruit in America—and in many other countries as well.

A few years ago my husband, Karl, and I moved from Nantucket Island to our earthly paradise in Key West, and we became fascinated with the gardening possibilities. We decided to try growing bananas in *our* Garden of Eden. Like some kind of exotic, tropical miracle the plantings worked and we had bananas—without labels! This fruitful abundance in our own backyard was the genesis of this cookbook. That first stem of bananas, all one hundred–plus of them ripening in unison was sublime inspiration.

My wish is that each reader of this book will someday enjoy the pleasure of having a banana plant within arm's reach. There's nothing quite like the flavor and light sweetness obtained from home-grown fruit. But supermarket bananas will do just fine, too, and will work in all the recipes that follow. I hope that they will fire your imagination to enjoy nature's perfect fruit in dozens of new ways.

Carol Lindquist
Love Lane
Key West, Florida
February 1992

1

Had Your Banana Today?

Bananas are indeed the wonder fruit. They come in easy to peel, biodegradable wrappings, and you can buy them (very inexpensively) all year round at virtually every food store, from the smallest corner grocery or fruit stand to the largest supermarket. They're handy and portable and quick to eat, and are amazingly versatile in the kitchen where they liven up everything from appetizers to elegant entrées to baked goods and desserts.

Bananas are also hard to beat when it comes to nutrition. They pack a walloping punch of potassium (important for proper muscle function and body-fluid balance) and vitamin B-6 (important for protein metabolism), as well as a host of other nutrients (see chart on page 3). In fact,*The New England Journal of Medicine* reported a 12–year scientific research study indicating that one extra serving per day of a potassium-rich food can cut the risk of death from stroke by as much as 40 percent. In addition, because of their high fiber content (bananas contain nonsoluble fiber and pectin), bananas provide a good source of much needed dietary bulk.

Is it any wonder that Americans consume about 11.5 billion bananas a year? On average, that's twenty-six pounds or approximately seventy-eight bananas per person, making it the number-one favorite fruit in America! Does all this sound too good to be fun? Well, read on and go bananas!

2

Bananas: The Nutritional Powerhouse

The following nutritional breakdown, supplied by the United States Department of Agriculture, is for a single, 100-gram (3.5-ounce) peeled banana. This would be an average-size banana of about 7–8 inches long.

Nutrient	% of U.S. Recommended Daily Allowance (RDA)
Vitamin B-6	29%
Vitamin C	15%
Vitamin A	*
Folacin	5%
Thiamine	3%
Riboflavin	5%
Niacin	3%
Magnesium	7%
Copper	6%
Iron	2%
Phosphorus	2%
Zinc	*
Calcium	*
Potassium	368 mg (no RDA is established)
Sodium	1 mg (no RDA is established)
Complex carbohydrates and natural sugars	23 mg (no RDA is established)
Fiber	2.1 g (no RDA is established)
Fat	.5 g (The banana 99.5% fat free.)
Calories	90

* Contains less than 2% of the U.S. RDA for this nutrient.

Tips on Buying and Storing Your Bananas

••• Bananas are one of the few fruits that are picked full size but green, and stored for a considerable time without injury to flavor. They must be ripened off the plant; bananas that are allowed to ripen on the plant tend to be mealy in texture.

••• Hang home grown bananas upside down in a cool, shady place to ripen slowly.

••• Green or semi-green bananas will ripen in a day or two at room temperature. For quicker ripening, place bananas in a paper bag overnight. For even faster ripening, add an apple to the bag.

••• Bananas need to breathe, so don't store them in a plastic bag— always use a paper bag.

••• To keep bananas from over-ripening, refrigerate them and use them within a couple of days. The skins will darken but the fruit will keep its texture, firmness, and flavor. Green bananas will be damaged by ordinary refrigeration. Once the ripening process has been interrupted by cold, the banana does not resume normal ripening when the temperature is raised.

•••Bananas can be safely frozen once they are ripe. In fact, frozen bananas are the basic ingredient in several delectable desserts (see pages 81–93). To freeze them, peel and dip whole bananas in lemon or lime juice to prevent discoloration. Wrap individual bananas snugly in plastic wrap so they are covered completely, with no air space. Once frozen, they keep for several months.

•••To prevent discoloration of peeled bananas that will not be served immediately, sprinkle them with lemon, lime, orange, or pineapple juice.

4

Banana Equivalents

2 to 3 medium, peeled bananas = 1 cup

4 to 5 medium, peeled bananas = 2 cups

1 pound = 3 medium bananas or 2+ cups sliced

1 pound = approximately 10½ ounces peeled fruit

Sliced: 1 medium banana = 30 slices ⅛-inch thick

Diced: 1 pound peeled = 2½ cups

Mashed: 1 medium banana = ⅓ to ½ cups

Breakfast & Brunch

Monkey's Delight

Serving a banana can be as simple as peel and eat—anytime, anywhere. Here are some suggestions for adding a bit of sophistication—and even a touch of royal class—to the peel-and-eat technique.

Drunken Dunkin' Banana

Legend has it that Walt Whitman was fond of bananas and, prior to each bite, would dip his banana into a glass of sherry. He might have tried it with rum as well.

Banana à la Windsor

The Duke suggested, "At breakfast time peel a banana. Lay it on a plate of ample size and ladle over it thick orange marmalade. Eat with a spoon, munching at the same time, hot buttered toast. Sip coffee."

All-American Classic

Sliced bananas on cornflakes with milk, of course!

Tami's Best Hampton Banana

Slice a large, ripe banana into ½ cup sour cream. Sprinkle with brown sugar or cover with a heaping spoonful of strawberry preserves.

Mark's Night-Before-Breakfast Banana

On the night before breakfast, mix ½ cup heavy cream and 2 tablespoons dark rum in a bowl. Slice a large, ripe banana into the mixture and "marinate" overnight. Sprinkle lightly with cinnamon and nutmeg before enjoying this sensational breakfast treat.

Super Banana Breakfast Shake

This is a quick and satisfying breakfast, but would also serve as a deliciously different lunch or snack.

> 2 cups skim milk
> 2 tablespoons honey
> 3 ripe bananas, cut into chunks
> 2 graham crackers, broken into pieces
>   pinch of nutmeg

1. In the container of a blender combine all ingredients.
2. Cover and blend at high speed for one minute.

Yield: 2 servings

For a thicker shake, substitute 1 cup nonfat yogurt for 1 cup skim milk.

A single plant bears only one stem of fruit. After it is harvested the plant is cut in order to allow the follower shoots, known as "daughter" and "granddaughter," to flourish.

Banana-Soufflé Pancake

This simple, delicious pancake has its origin in Yorkshire Pudding.

½ cup milk
2 eggs
½ cup flour
  pinch of nutmeg
¼ cup butter
1 large banana, cut crosswise
    into ¼-inch slices
3 tablespoons confectioner's sugar
3 tablespoons Grand Marnier liqueur

1. Preheat over to 425° F.
2 In a medium-size bowl beat milk and eggs together with a fork until well mixed.
3. Add flour and nutmeg and mix with a fork until just combined.
4. In a 10-inch skillet with an oven-proof handle, melt butter over a moderate heat until very hot, being careful not to burn.
5. Remove skillet from heat, pour in batter, and arrange banana slices on top.
6. Bake for 15 minutes. When pancake is puffed up, remove skillet from oven.
7. Sprinkle confectioners sugar over top of pancake, then pour the Grand Marnier over all.
8. Return to oven for one minute.
9. To serve: Cut into halves or quarters, depending on desired serving size.

Yield: 2 to 4 servings

Pancakes with Ginger-Maple Sauce

Making pancakes from scratch is nearly as easy as using a mix, and the taste is infinitely better.

Pancakes	
2	cups flour
3	teaspoons baking powder
	pinch of salt
	pinch of nutmeg
2	large ripe bananas, finely diced
1½	cups milk
1	teaspoon vanilla
1	egg
2	tablespoons vegetable oil

1. Combine flour, baking powder, salt, and nutmeg in a large mixing bowl.
2. Gently toss in diced bananas and coat with dry mixture.
3. In a smaller bowl beat milk, vanilla, egg, and oil until well blended. Pour liquid mixture into bowl with dry ingredients and stir only enough to moisten.
4. Spoon batter onto a lightly greased, hot griddle. When batter begins to bubble, turn pancakes and cook until second side is lightly browned.

Sauce	
¼	cup butter
½	cup maple syrup
¼	cup ginger conserve

1. Melt butter in a small saucepan over low heat.
2. Stir in maple syrup and ginger conserve and continue to cook over a low heat, stirring until well blended and hot.
3. Spoon over pancakes.

Yield: Approximately two dozen 4-inch pancakes

10

Sunday Scalloped Bananas

This dish is ideal as the main course for brunch or as a luncheon dessert.

2	cups soft bread crumbs
2	tablespoons melted butter
¼	cup golden raisins
2	large bananas, cut crosswise into ¼-inch slices
½	cup sugar
	zest and juice of half an orange
½	teaspoon cinnamon
¼	teaspoon nutmeg

1. Preheat oven to 350° F.
2. Grease an 8 x 8 x 1½ inch baking pan.
3. In a medium-size mixing bowl combine crumbs, butter, and raisins. Set aside.
4 In a medium-size mixing bowl combine remaining ingredients.
5. Place alternate layers of the two mixtures in prepared baking pan, starting with banana mixture and ending with crumb mixture.
6. Bake for 15 minutes. Remove pan from oven and cool slightly before serving.
   NOTE: This dish is good with a little half & half poured over it.

Yield: 4 servings

Grilled Banana-Bacon Roll-Ups

Banana and bacon have a great natural flavor affinity. These taste best when cooked on a grill but a broiler will do as well. Serve with warm cornbread for a nice company brunch dish.

1	cup orange juice
¼	cup dark rum
4	large ripe bananas
¼	cup brown sugar
4	slices bacon

1. The night before serving, or at least one hour before, in a shallow dish large enough to hold the bananas, combine orange juice and rum. Place the whole peeled bananas in this mixture and allow to marinate, turning occasionally.
2. When ready to cook, remove bananas from marinade, allowing any excess to drip off.
3. Roll bananas in crumbled brown sugar.
4. Wrap bacon in spiral fashion lengthwise around bananas. Secure with a toothpick if necessary.
5. Place on grill and cook while turning gently and frequently until bacon is crisp. If a grill is not convenient, place bananas on a broiler pan and broil until bacon is crisp on one side, then turn and broil until crisp on other side.

Yield: 4 servings

Banana Frittata

This makes a nice luncheon dish especially when served with a salad and biscuits.

2 tablespoons butter
4 pre-cooked "brown-and-serve" sausage links cut into ½-inch pieces
2 bananas, cut crosswise into 1-inch slices
1 tablespoon honey mustard
½ cup crushed pineapple, well drained
6 eggs, slightly beaten

1. Preheat oven to 450° F.
2. In a 9-inch Pyrex glass pie plate place butter and sausage and bake for 4 minutes.
3. Remove from oven and reduce heat to 400° F. Arrange banana slices in with the sausage.
4. Stir honey mustard into crushed pineapple and place this mixture on top of sausage and bananas.
5. Pour beaten eggs over all and immediately return to oven. Bake approximately 20 minutes to set. Do not not overcook. Remove dish from oven and serve hot.

Yield: 2 to 4 servings

Fluffy Banana Omelette

The club soda and bananas make a plain egg omelette light, fluffy, and flavorful.

- 1 tablespoon butter
- 2 extra-large eggs
- 1 small, very ripe banana
- 1 tablespoon club soda

1. Melt butter in a 10-inch Teflon frying pan over low heat.
2. Break eggs into the container of an electric blender. Cover and blend on high speed for 30 seconds.
3. Break banana into chunks and add to mixture. Blend again for 30 seconds.
4. Stir in club soda.
5. Turn up heat under butter to medium-high and pour in egg mixture. Do not stir. When small bubbles appear on the surface, use a rubber spatula to carefully fold in half. Gently slide omelette onto serving platter and serve hot.

Yield: 2 servings

Bananas are not grown commercially in the United States, but we are the world's largest importer and Latin America is the largest exporter.

Banana-Mousseline Omelette

Try this omelette as a festive brunch or after-theater treat. This is my friend Pat's version of a favorite so many of us enjoyed at Clyde's Restaurant in Washington, D.C., during the days of Camelot.

4 large eggs, separated
¼ teaspooon cinnamon
2 tablespoons sugar
1 tablespoon butter
3 tablespoons brown sugar
1 ripe banana, cut into thin slices
2 tablespoons sour cream
confectioners' sugar

1. Turn on broiler.
2. In a large mixing bowl, and using an electric mixer, beat egg whites with 2 tablespoons sugar until stiff
3. In a medium bowl, and using an electric mixer, beat egg yolks and cinnamon until pale and thick.
4. Using a rubber spatula, carefully fold yolks into whites.
5. In a 7-inch omelette pan with a broiler-proof handle, heat butter until hot, but not smoking, over medium-high heat. Pour in egg mixture and gently spread around pan with rubber spatula. Cook for one minute.
6. Sprinkle with brown sugar and put under broiler to caramelize the sugar. Remove from broiler and arrange half the banana slices in center of omelette. Carefully fold omelette in half and slide it onto a serving plate.
7. Garnish with remaining banana slices and sour cream. Dust with sifted confectioners' sugar. Serve hot.

Yield: 2 servings

Muffins &
Quick Bread

Quick-and-Sticky Buns

Quick plus sticky equals easy and delicious. These buns offer truly instant gratification.

- 10 teaspoons butter
- 10 teaspoons brown sugar
- ½ cup pecan halves
- 2 bananas, sliced
- 1 10-ounce package refrigerated buttermilk biscuits

1. Preheat oven to 400° F.
2. Put 1 teaspoon butter into each of 10 muffin-pan cups. Place muffin pan in oven until butter melts.
3. Add 1 teaspoon brown sugar to each cup, then several pecan halves, and 3 slices banana. Cover with a biscuit, gently pressed down.
4. Bake for 10 minutes or until golden. Remove pan from oven, cool for 5 minutes on a wire rack, and then invert onto a large plate. Serve warm.

Yield: 10 buns

Hindu legend says the banana was the forbidden fruit in the earthly paradise and it was with the leaves of this plant that the first man and woman covered their nakedness.

Banana–Raisin Bran Muffins

Each of these tasty, nutrition-packed muffins is a breakfast in itself.

½	cup orange juice
2	eggs
2	large ripe bananas, cut in chunks
¼	cup butter, softened to room temperature
⅔	cup light brown sugar
1	teaspoon vanilla
1	cup all-purpose flour
1	cup raisin bran flakes
2	teaspoons baking powder
1	teaspoon salt
1	teaspoon ground allspice

1. Preheat oven to 375° F. Grease or line muffin pans with cupcake papers.
2. Place orange juice, eggs, bananas, butter, brown sugar, and vanilla in the container of an electric blender. cover and blend on high speed for 30 seconds.
3. In a large mixing bowl combine flour, raisin bran, baking powder, salt, and allspice. Pour liquid mixture over this and mix only enough to moisten dry ingredients.
4. Fill prepared muffin pan ⅔ full and bake for 25 to 30 minutes. Remove pan from oven and allow to cool on a wire rack.

Yield: 12 muffins

Key West–Nantucket Muffins

When the bananas of Key West meet the blueberries of Nantucket you have a gastronomic marriage made in heaven.

        ½ cup butter, softened to room temperature
        1 cup sugar
        1 egg
        ½ cup buttermilk
        1 teaspoon lemon zest
        2 large ripe bananas, mashed
       2¼ cups all-purpose flour
        2 teaspoons baking powder
        ½ teaspoon baking soda
        ½ teaspoon salt
        ¼ teaspoon nutmeg
        1 cup blueberries

1. Preheat oven to 350° F. Grease or line with cupcake papers a 12-inch muffin pan.
2. In a large mixing bowl and using an electric mixer, cream butter and sugar together. Add egg and buttermilk and beat until fluffy. Add lemon zest and bananas, beating to combine.
3. In a medium-size bowl, combine flour, baking powder, baking soda, salt, and nutmeg. Gradually add remaining dry ingredients to banana mixture, beating after each addition to blend completely.
4. Gently fold in blueberries.
5. Spoon into prepared muffin pan so that each cup is ⅔ full. Bake for 25 to 30 minutes. Remove pan from oven and allow to cool on wire rack.

Banana-Twinkie Muffins

These stir-and-stuff muffins are easy, delicious, and nutritious—a winning combination.

½ cup milk
½ cup orange juice
¼ cup vegetable oil
1 egg
1 cup bran flakes
1 cup rolled oats
½ cup sugar
½ cup brown sugar
1¼ cups flour
1 tablespoon baking powder
½ teaspoon baking soda
½ teaspoon salt
2 medium bananas, each cut into 6 pieces

1. Preheat oven to 350° F. Grease or line with cupcake papers a 12-inch muffin pan.
2. In a large mixing bowl combine milk, juice, oil, and egg. Beat with a fork to blend. Add bran flakes, oats, and sugars. Stir well with a fork to mix. Set aside for 5 minutes so that bran flakes and oats absorb liquid.
3. Add flour, baking powder, baking soda, and salt, stirring well with a fork to blend all ingredients.
4. Spoon into prepared pan. Submerge one chunk of banana into each muffin.
5. Bake for 20 to 25 minutes or until golden. Remove pan from oven and allow to cool on a wire rack.

Yield: 12 muffins

Love Lane Banana Bread

This bread is especially good for tea sandwiches. Slice thinly and spread with softened cream cheese.

½ cup butter, softened to room temperature
1 cup sugar
2 eggs
zest and juice of 1 orange
4 ripe bananas, coarsely mashed
2 cups all-purpose flour
1 teaspoon baking soda
½ teaspoon salt
½ teaspoon nutmeg
¼ teaspoon ground ginger

1. Preheat oven to 350° F. Grease and flour a 9 x 5 inch loaf pan.
2. In a large mixing bowl and using an electric mixer, cream butter and sugar together. Add eggs and beat until fluffy. Add orange zest, juice, and bananas, beating to combine.
3. Add remaining dry ingredients and beat just enough to incorporate completely.
4. Pour into prepared pan and bake for 1 hour or until wooden toothpick inserted in center of loaf comes out clean. Remove pan from oven and allow to cool in the pan for 10 minutes. Invert onto wire rack to complete cooling.

Yield: 1 loaf

FOR VARIETY: Add one of the following—1 cup coarsely chopped walnuts, black walnuts, peanuts, pecans, finely chopped dates, golden raisins, or shredded coconut.

Banana-Apricot Tea Loaf

For a special treat this may be sliced and lightly toasted and buttered or spread with Orange Cream Cheese (see page 24).

1	cup all-purpose flour
¾	cup whole wheat flour
¼	teaspoon ground cardamon
2¼	teaspoons baking powder
½	teaspoon salt
⅓	cup butter, softened to room temperature
⅔	cup sugar
1	teaspoon lemon zest
2	eggs
3	ripe bananas
½	cup pecan
¼	cup dried apricots, finely chopped confectioners' sugar

1. Preheat oven to 350°F. Grease and flour an 8½ x 4½ inch loaf pan.
2. In a medium mixing bowl sift together flours, cardamon, baking powder, and salt.
3. In a large mixing bowl and using an electric mixer, cream butter, sugar, and lemon zest. Beat in eggs and bananas. Gradually add dry ingredients, beating after each addition to blend completely. Stir in nuts and apricots by hand.
4. Pour into prepared pan and bake for 1 hour or until a wooden toothpick inserted in center of loaf comes out clean. Remove pan from oven and allow to cool in the pan for 10 minutes. Invert onto wire rack to complete cooling. When cool, dust top with sifted confectioners' sugar.

Yield: 1 loaf

Mile Zero Banana Cornbread

Mile Zero is another name for Key West, Florida. It also marks the beginning of U.S. Route 1, in Key West, and is immortalized by the novel of the same name by Thomas Sanchez.

1	cup yellow cornmeal
1	cup all-purpose flour
2	teaspoons baking powder
½	teaspoon baking soda
¼	cup sugar
½	teaspoon ground cumin
1	teaspoon salt
2	ripe bananas, coarsely mashed
½	cup milk
1	egg, beaten
½	cup plain yogurt
1	small can chopped green chilies

1. Preheat oven to 375° F. and grease an 8 x 8 inch pan.
2. In a small mixing bowl, combine all dry ingredients.
3. In a large mixing bowl and using an electric mixer, combine bananas, milk, egg, yogurt, and chilies, beating to mix. Add dry ingredients to banana mixture and mix only to combine. Do not overmix.
4. Pour into prepared pan and bake for 25 to 30 minutes, or until golden. Remove pan from oven and serve hot with soup, chicken salad, chili, or scrambled eggs for a hearty lunch or Sunday supper.

Yield: 9 servings

Banana-Pumpkin Spice Loaf

This spicy loaf is good with all that leftover Thanksgiving turkey.

½ cup butter, softened
   to room temperature
½ cup sugar
½ cup brown sugar
2 eggs
3 tablespoons frozen orange
   juice concentrate, thawed
1 tablespoon grated fresh ginger
2 ripe bananas, mashed

½ cup pumpkin purée
2½ cups all-purpose flour
1 teaspoon cinnamon
⅛ teaspoon cloves
1 teaspoon baking powder
½ teaspoon baking soda
½ cup pecan pieces
½ cup golden raisins

1. Preheat oven to 350° F. Grease and flour a 8½ x 4½ inch loaf pan.
2. In a large mixing bowl and using an electric mixer, cream butter and sugars. Add eggs and beat until fluffy. Beat in orange juice concentrate, ginger, banana, and pumpkin. Gradually add flour, spices, baking powder, and soda, beating after each addition to blend completely. Stir in pecans and raisins by hand.
3. Pour batter into the prepared pan. Bake for approximately 1¼ hours or until a wooden toothpick inserted in the center of the loaf comes out clean. Remove pan from oven and allow to cool for 10 minutes in pan. Invert onto wire rack to complete cooling.

Yield: 1 loaf

For delicious tea sandwiches, spread Orange Cream Cheese onto thin slices of the Banana-Pumpkin Spice Loaf .

To make Orange Cream Cheese combine one 8-ounce package cream cheese (softened to room temperature), 3 tablespoons frozen orange juice concentrate (thawed), and 1 tablespoon orange zest, and beat until blended and smooth.

24

Appetizers, Condiments & Dressings

Easy Banana-Tomato Chutney

This is a quick-and-easy, no-cook chutney. It's excellent with grilled fish, lamb, or chicken.

2 ripe bananas, diced
2 ripe tomatoes, diced and drained
1 tablespoon Tiger brand hot pepper sauce
½ teaspoon worcestershire sauce

1. In a small mixing bowl combine all ingredients.
2. Chill well before serving.

Yield: approximately 2 cups

Call (818) 798-2272 for the following recorded message from Ken Bannister, a.k.a Bananister, Top Banana and founder of The International Banana Club: "This is Ken Bannister, welcome to the bunch!" Created in 1972 and dedicated "to keep people smiling," the club has a membership fee of $10 which includes a gold-plated banana-shaped pen, a sheet of small stickers that say "one of the bunch," a newsletter that is published "irregularly" and admission to the world's only Banana Museum, located in Altadena, California. It houses over 15,000 pieces of "bananabilia" including a petrified banana left in a closet for over five years.

Karl's Cayo Hueso B & B Dip

This hearty dip incorporates two favorite Cuban food staples used in Key West (Cayo Hueso), Florida.

    1 15-ounce can Kirby brand
           seasoned black beans*
    1 large ripe banana
    1 tablespoon lime juice
    2 tablespoons Tiger brand hot pepper

1. Drain liquid from beans and set aside.
2. In a medium-size saucepan mix all ingredients and mash with a potato masher to a smooth consistency, adding some of the reserved liquid if necessary.
3. Heat and serve with tortilla chips

Yield: 2 cups (enough for a group of 4 to 6 people)

*If using an unseasoned brand of black beans, add ½ teaspoon each of crushed dried bell pepper flakes, onion, and garlic powder; 1 teaspoon each of salt and sugar; and 1 tablespoon vinegar.

Banana Rumaki

This dish wins rave reviews every time I serve it. You can never make too many. It's the perfect finger food for a cocktail party.

½	cup soy sauce
½	cup orange juice
¼	cup packed brown sugar
1	clove garlic, crushed
½	teaspoon curry powder
¼	teaspoon ground ginger
¼	teaspoon cinnamon
4	bananas, cut into 1-inch slices
1	8-ounce can water chestnuts, cut into halves
12	slices lean bacon, halved crosswise
1	package wooden toothpicks

1. In a broad, shallow bowl combine soy sauce, orange juice, brown sugar, garlic, curry, ginger, and cinnamon. Stir well.
2. Wrap ½ slice bacon around one slice banana and ½ water chestnut. Fasten with a wooden toothpick. Repeat, until you have 24 rumaki.
3. Marinate in sauce at least 1 hour, turning occasionally.
4. Preheat oven to 400° F.
5. Remove rumaki from marinade and place on a large, shallow baking sheet.
6. Bake 12 to 15 minutes until bacon is crisp. Remove rumaki from baking sheet and drain on paper towel. Serve warm.

Yield: About 24 pieces

Latitude 24 Banana Tidbits

These hot hors d'oeuvres are easy enough for even the most laid-back Chiquita. If you have the ingredients on hand you can make this any time unexpected company drops by or for a last-minute get-together.

½ cup chili sauce
1 cup grape jelly
2 tablespoons lime juice
2 tablespoons Tiger brand hot pepper sauce
2 tablespoons brown sugar
¼ teaspoon ground cloves
4 large green-tipped bananas,
     cut into 1-inch slices

1. In a heat-proof serving or chafing dish combine all ingredients except bananas, mixing and heating thoroughly over medium heat. Add bananas, stirring gently to coat. Serve with wooden toothpicks.

Yield: About 32 pieces

Banana Crackers

Created by my husband, Karl, these small crackers are good enough to eat plain, or spread with peanut butter or cream cheese. They make an interesting base for your favorite party hors d'oeuvres.

¼ cup butter
½ cup all-purpose flour
⅛ teaspoon salt
4 drops Tabasco sauce
1 ripe banana, mashed
   coarse salt

1. Preheat oven to 450° F. and grease a 12 x 15 inch cookie sheet.
2. In a medium-size mixing bowand using a pastry blender, mix butter, flour, and salt together until crumbly. Using a fork, stir in Tabasco and banana.
3. On a floured surface, knead mixture gently, adding a small amount of flour if necessary to make a firm dough.
4. Divide dough in half and roll out one half at a time on a floured surface until thin enough to cover prepared cookie sheet.
5. Place dough on cookie sheet and sprinkle lightly with coarse salt.
6. Using a pastry cutter, cut dough into 1-inch squares. Bake 6 to 8 minutes or until lightly browned. Remove crackers with a spatula and place on a wire rack to cool. Repeat with second half of dough.

Yield: 30 dozen crackers

Jamaican Banana Jam

For a change of pace from the familiar berry jams, give your peanut butter sandwich or your breakfast toast a tropical flavor. The recipe yields enough to fill six 8-ounce jars and makes the perfect hostess gift.

> 1   cup orange juice
> ½   cup lime juice
> ½   cup water
> 2   cups sugar
> 5   large ripe bananas, thinly sliced

1. In a large, heavy saucepan combine all ingredients, except bananas, and bring to a boil over medium heat.
2. Add bananas and lower heat to a simmer, stirring regularly until mixture is thick.
3. Ladle into hot, sterilized jars and seal immediately according to instructions accompanying your particular type of canning jars.

Yield: Six 8-ounce jars

The Chiquita Banana song was created in 1942 as a radio advertising jingle and quickly became a jukebox hit that was played 376 times a day at its peak.

Three Great Appetizer Combinations

Bananas combine well with other ingredients to make interesting appetizers. Here are three of my favorites.

1. Wrap thinly sliced prosciutto ham around chunks of bananas that have been dipped in lime juice. Fasten ham with a wooden toothpick.

2. Coat diagonally cut banana slices with finely chopped nuts. Top with a small amount of one of the following:
- Chutney
- Peanut butter with a drop of Tiger brand hot pepper sauce
- Deviled ham mixed with sweet pepper relish
- Ginger-flavored cream cheese

3. Cut several bananas into one-inch slices and scoop out a hollow in each slice with a melon baller. Dip slices in lemon or lime juice to prevent discoloration. Fill with cream cheese seasoned with one of the following: herbs, garlic, dill, chives, or mint. Use the banana balls in a fresh fruit salad.

According to the Guiness Book of World Records, *the longest banana split ever made was 4.55 miles in length and created by the residents of Selinsgrove, Pennsylvania.*

Maggie's Swedish Banana Relish

Maggie is a Swedish weaver living on Nantucket Island, whose cooking is as inventive as her weaving. She likes to use this piquant relish with grilled fish or chicken.

1 large ripe banana, mashed
2 apples, peeled and grated
2 small carrots, peeled and grated
3 tablespoons mayonnaise
  juice of 1 lemon
1 tablespoon milk
1 clove garlic, crushed
¼ teaspoon white pepper
¼ teaspoon salt
1 teaspoon sugar
2 tablespoons chopped fresh dill

1. In a mixing bowl combine all ingredients and mix thoroughly.
2. Chill and serve very cold.

Yield: approximately 1½ cups

John's Banana-Tamarind Chutney

The tamarind is a tropical fruit with a somewhat bitter and astringent, yet refreshing, flavor. The pulp is available frozen in some supermarkets or in specialty food stores, especially those selling Indian foods.

Our friend John is an eclectic cook. This is one of his favorite chutney recipes. It is a no-cook chutney, and is best when eaten as soon as it's ready. It is excellent with curries and roasted lamb. Combine it with cream cheese for an unusually tasty sandwich.

2	tablespoons lemon juice
1	large ripe banana, diced
½	cup tamarind pulp
1	tablespoon golden raisins
1	teaspoon ground cumin
¼	teaspoon cayenne pepper
1	teaspoon salt
2	tablespoons sugar

1. In a small glass or ceramic mixing bowl sprinkle lemon juice over diced banana.
2. Add remaining ingredients and stir well with a fork to mix

Yield: 1 cup

Banana–Poppy Seed Dressing

All of the following savory and sweet dressings are virtually fat and cholesterol free, and make excellent dips for fruit and vegetables. This first one is particularly good on chicken salad, fruit salad, or grilled fish.

1 large (or 2 small) ripe bananas cut into chunks
1 cup plain nonfat yogurt
¼ cup light brown sugar
2 tablespoons orange juice
1 tablespoon raspberry or cider vinegar
¼ teaspoon ground ginger
1 tablespoon poppy seeds

1. Place all ingredients except poppy seeds in container of an electric blender.
2. Cover and blend on high speed until smooth.
3. Pour dressing into sealable container or jar, add poppy seeds, and Refrigerate until ready to use.

Yield: Approximately 1½ cups

Sherry–Honey Mustard Dressing

This is good on grilled chicken or fish or can be used as a savory dip for vegetables.

    1   large (or 2 small) ripe bananas
          cut into chunks
    1   cup plain nonfat yogurt
    2   tablespoons honey mustard
    2   tablespoons sherry vinegar

1. Place all ingredients in container of an electric blender.
2. Cover and blend on high speed until smooth.
3. Pour dressing into sealable container or jar and refrigerate until ready to use.

Yield: Approximately 1½ cups

The ragged-edge of the giant banana leaf is nature's defense against heavy tropical winds and rain.

Curry Dressing

This is good on grilled chicken or fish, or can be used as a savory dip for vegetables.

1 large (or 2 small) ripe bananas cut into chunks
1 cup plain nonfat yogurt
2 tablespoons orange juice
2 tablespoons light brown sugar
1 tablespoon cider vinegar
¼ teaspoon ground ginger
1 teaspoon curry powder

1. Place all ingredients in container of an electric blender.
2. Cover and blend on high speed until smooth.
3. Pour dressing into sealable container or jar until ready to serve.

Yield: Approximately 1½ cups

Bananas debuted officially in America in 1876 at Philadelphia's Centennial Exhibition. It was wrapped in colored tinfoil—exotic and costing 10 cents. Today the banana is America's most popular fruit and not much more expensive.

Soups, Salads & Sandwiches

West Indies Hot Banana-Peanut Soup

My husband and I first experienced this soup while sailing in the Windward Islands, where bananas, hot peppers, and peanuts are abundant, and where each island makes its own hot pepper sauces. This recipe is my version, which is the delicious and inevitable consequence of combining these readily available ingredients.

      2 cups chicken broth
      ½ cup chunky peanut butter
      3 large, ripe bananas, coarsely mashed
 1 to 2 tablespoons Tiger brand hot pepper sauce
      ½ cup toasted coconut

1. In a medium-size heavy saucepan heat chicken broth over medium heat and gradually stir in peanut butter until well blended.
2. Add mashed bananas and hot pepper sauce to taste.
3. Pour into individual serving bowls, sprinkle with toasted coconut, and serve hot.

Yield: 4 servings

The current banana eating record is sixteen 3-ounce bananas in 2 minutes and was set at the Banana Pig-Out, sponsored by the International Banana Club.

Banana Gazpacho Tropicale

This fruitful variation on the traditional Spanish cold tomato gazpacho soup makes an ideal first course for a hot summer night's dinner. It's also perfect as the main course for a summer luncheon.

1	cup blanched almonds
1	cup orange juice
4	large, ripe bananas, in chunks
1	cup strawberries, hulled and halved
1	cup plain yogurt
¼	cup raspberry vinegar
1	teaspoon ground ginger
1	cup apricot nectar
1	cup pineapple nectar

1. Using a blender or food processor with a steel blade, process almonds and orange juice until smooth.
2. Add bananas, strawberries, yogurt, raspberry vinegar, and ginger and continue to process until thoroughly blended.
3. Pour mixture into a large mixing bowl and stir in apricot and pineapple nectars. Mix thoroughly.
4. Refrigerate for several hours. Serve well chilled. Garnish with strawberry halves or mint leaves if desired.

Yield: 8 servings

Banana Waldorf Salad

Try this old-fashioned favorite with a tropical twist.

       ¼ cup orange juice
       2 large bananas, sliced ¼-inch thick
       2 large apples, cored and coarsely chopped
       ½ cup thinly sliced celery
       ½ cup coarsely broken walnuts
  ¼ to ½ cup Banana–Poppy Seed Dressing (page 35)

1. In a large mixing bowl sprinkle orange juice over banana slices.
2. Add remaining ingredients, using enough Banana–Poppy Seed Dressing to moisten, tossing the mixture lightly.
3. Chill thoroughly and serve on crisp salad greens.

Yield: 2 servings

*In the tropical jungles of Malaysia where the banana was
born, there are wild varieties with hard pits, untouched by
the evolution of seedless varieties.*

Last Resort Fruit Salad

The Banana–Poppy Seed Dressing on page 35 is a tasty addition to this salad.

2	tablespoons unflavored gelatin
¼	cup cold water
½	cup boiling water
¼	cup lemon juice
2	tablespoons sugar
1	cup ginger ale
2	bananas, sliced
1	cup seedless white grapes
½	cup chopped pecans

1. Soak gelatin in cold water for 5 minutes. Place in a large mixing bowl and add boiling water to dissolve gelatin. Add lemon juice, sugar, and ginger ale. Stir well to mix.
2. Refrigerate until syrupy, approximately 45 minutes.
3. Fold in sliced bananas, grapes, and pecans.
4. Pour into a one-quart ring mold and chill until well set, at least 1 hour.
5. Unmold onto a chilled serving platter. Surround with crisp lettuce leaves.

Yield: 4 servings

Trade Winds Coleslaw

3 bananas, coarsely diced
1 8-ounce can pineapple tidbits
4 cups shredded cabbage
1 large carrot, grated
2 tablespoons cider vinegar
½ cup mayonnaise
¼ teaspoon ground ginger

1. In a large mixing bowl combine bananas and pineapple tidbits and their juice.
2. Add remaining ingredients and toss lightly to combine well. Chill.

Yield: 6 servings

You can use the Banana–Poppy Seed Dressing (page 35) as a substitute for the mayonnaise and vinegar, in which case you will need about ½ cup dressing.

Charlie Chaplin, in his poor and struggling youth, survived for one month on a case of bananas.

In 1924, Josephine Baker danced in Paris at the Folies-Bergère with a string of bananas around her waist.

Tropical Fruit Salad

A refreshing summertime lunch.

      4 large bananas, cut in half lengthwise
      1 14-ounce can sweetened condensed milk
     ¾ cup finely chopped peanuts or pecans
      1 8-ounce can pineapple tidbits (drain and
           reserve juice)
      1 11-ounce can mandarin oranges (drain and
           reserve juice)
     ½ cup mayonnaise
         iceberg lettuce

1. Carefully dip banana halves in sweetened, condensed milk and then roll in chopped nuts to coat.
2. On a bed of iceberg lettuce, arrange 2 banana halves to form a circle.
3. In a small mixing bowl carefully combine pineapple tidbits and mandarin oranges. Spoon ¼ of the mixture into center of banana circle.
4. Thin mayonnaise to desired consistency with some of reserved fruit juices and drizzle over salad.

Yield: 4 servings

Grenadine Banana Salad

The grenadine syrup colors the sour cream a pretty shade of pink as well as imparting a nice flavor.

> 3 bananas, sliced half-inch thick
> 1 small cantaloupe, peeled, seeded,
>   and cut into balls with a melon baller
> 1 cup strawberries, quartered
> ½ cup orange juice
> 1 cup sour cream
> ¼ cup grenadine syrup
>   mint sprigs for garnish

1. In a large mixing bowl combine fruit and orange juice.
2. Divide equally among six dessert bowls.
3. In a small mixing bowl combine sour cream and grenadine syrup. Using a fork, mix well. Spoon over the fruit and garnish each serving with a sprig of mint.

Yield: 6 servings

Bananas are 99.5 % fat free—ounce for ounce—on a par with lettuce.

Banana–Grand Marnier Compote

This fruit salad is an excellent accompaniment to roasted pork as well as an appealing dessert or an elegant breakfast treat.

½ cup orange juice
½ cup water
1 cup sugar
zest of 1 orange
2 tablespoons Grand Marnier
6 large bananas, cut into half-inch slices

1. In a medium-size saucepan over medium heat combine orange juice, water, sugar, and zest. Bring to a boil and simmer for 5 minutes, stirring occasionally. Remove pan from heat, add Grand Marnier, and cool slightly.
2. Place banana slices in a serving bowl and pour on warm syrup. Stir gently to coat. Allow to macerate at room temperature for 20 minutes. Serve at room temperature or serve chilled.

Yield: 6 to 8 servings

Banana Sandwich Combinations

Bananas add an interesting flavor and texture to some old favorites:

1. Banana slices and chutney (see page 26) on date-nut bread.

2. Banana slices, bacon, and peanut butter on whole wheat bread.

3. Banana slices and orange marmalade or ginger conserve on cinnamon-raisin bread.

4. Banana slices, thinly sliced ham, and honey mustard on rye bread.

5. Banana and avocado slices, sliced smoked turkey breast, and mayonnaise on pumpernickel bread.

6. Mashed banana with chopped pecans and well-drained, crushed pineapple on oatmeal bread.

Hearty & Savory Dishes

Banana Ham Loaf

This makes a nice luncheon dish served cold.

  1 egg
  ½ cup apple cider
  1 teaspoon coriander
  3 tablespoons green tomato piccalilli relish or dill
       pickle relish
  1 pound coarsely ground lean ham
  1 cup cornbread stuffing mix
  2 large bananas
  ¼ cup honey mustard

1. Preheat oven to 325° F. Lightly oil a 9 x 5 inch loaf pan.
2. In a large mixing bowl and using a fork, beat egg, cider, coriander, and relish until blended. Add ham and cornbread mix. Stir with fork to mix completely. Divide in half.
3. Pack one half into prepared loaf pan. Place whole peeled bananas side by side on top. Add remaining ham mixture and press down gently. Spread with honey mustard.
4. Bake approximately 45 minutes or until ham is bubbly and slightly browned. Remove pan from oven and allow to cool slightly before slicing.

Yield: 6 to 8 servings

Stuffed Chicken Breasts

Your luncheon or dinner guests will love this tropical dish with Oriental overtones.

- 4 chicken breasts, boned and left whole with the skin on, patted dry
- 8 tablespoons chunky peanut butter
- 4 scallions, white part only
- 4 teaspoons freshly grated ginger
- 2 small bananas, halved lengthwise
- 4 tablespoons melted butter
- salt, pepper, and paprika to taste

1. Preheat oven to 350° F. Lightly butter a shallow pan just large enough to comfortably hold the 4 chicken breasts.
2. Spread 2 tablespoons peanut butter over boned side of each chicken breast. Place 1 scallion and 1 teaspoon ginger on each. Top each breast with a banana half.
3. Fold top and bottom of each chicken breast over about 1 inch, then roll up breast from the side.
4. Place each stuffed breast, seam side down, in prepared pan. Brush breasts with melted butter and sprinkle lightly with salt, pepper, and paprika.
5. Bake for 50 to 55 minutes until golden brown.

Yield: 4 servings

Plantation Casserole

This variation on the Haitian original makes an excellent accompaniment to pork or poultry.

      4 large sweet potatoes, peeled, cut into chunks,
           and boiled until tender
      3 large bananas
      ¼ cup butter
      2 tablespoons cream
      2 tablespoons dark rum
      ¼ cup loosely packed brown sugar
      ¾ cup drained crushed pineapple
      1 teaspoon salt
      ¼ teaspoon ground nutmeg
      ½ cup pecan pieces

1. Preheat oven to 350° F. Grease a 2½-quart ovenproof casserole.
2. Place potatoes, bananas, butter, cream, rum, and sugar in container of a food processor fitted with a steel blade. Process until smooth. Stir in pineapple, salt, and nutmeg.
3. Pour into prepared casserole and sprinkle with pecan pieces. Dot with some butter. Bake for 25 minutes. Serve hot.

Yield: 8 servings

Fruitful Roasted Pork

This method of roasting results in a succulent and tender dish with a wonderful blending of flavors.

 1 oven cooking bag used for roasting meat (available in most supermarkets)
 1 tablespoon all-purpose flour
 2 pounds boneless, rolled pork or pork loin
 4 small, firm bananas, whole
 1 cup dried apricots
 1 cup pitted prunes
 1 cup firmly packed brown sugar
 ½ cup orange juice

1. Preheat oven to 325° F. Dust inside of oven cooking bag with flour.
2. Place pork, bananas, and dried fruit in bag. Place bag in a large, shallow roasting pan.
3. Sprinkle brown sugar into bag and pour in orange juice. Tie bag securely and with a fork poke 4 holes evenly spaced across top of bag.
4. Roast for 1½ hours or until a meat thermometer inserted in center of roast registers 170° F.
5. Remove pan from oven, carefully slit open bag, and place meat on a serving platter. Arrange fruit around it. Spoon juices over meat and fruit.

Yield: 4 to 6 servings

Banana Fritters

In Florida and the Caribbean, plain banana fritters are served as a side dish with seafood, chicken, or pork. In Indonesia, these fritters traditionally accompany the Balinese main course. They also make a nice dessert sprinkled with powdered sugar or maple syrup.

> 1 cup all-purpose flour
> 1½ teaspoons baking powder
> 2 tablespoons sugar
> ½ teaspoon salt
> ½ cup milk
> 1 egg, well beaten
> 1 tablespoon dark rum
> 4 large ripe bananas, each cut into
>   4 diagonal pieces
> ¼ cup all-purpose flour for coating
>   vegetable oil for cooking

1. In a medium-size bowl mix together 1 cup sifted flour, baking powder, sugar, and salt. Add milk, well-beaten egg, and rum to dry ingredients and, using a whisk, mix until batter is smooth.
2. Roll banana pieces in ¼ cup flour. Shake off excess and dip each piece into batter, completely coating banana.
3. Fry in hot (375° F) deep oil for 4 to 6 minutes, turning to brown evenly. Remove and drain on a paper towel.

NOTE: Rolling the banana pieces in flour before dipping them in the batter makes them extra crispy.

Yield: 4 to 6 servings

Banana-Shrimp Curry

This is a terrific and festive company dish. Serve it with the following condiments in separate small bowls: shredded coconut, peanuts, chutney, chopped scallions, raisins, and pineapple tidbits.

½ cup butter divided
3 large, firm bananas, cut into 1-inch chunks
1 large onion, chopped
3 tablespoons all-purpose flour
1 cup applesauce
1 can condensed beef bouillon
1 tablespoon curry powder
¼ teaspoon ground ginger
dash of cayenne pepper
3 tablespoons lemon juice
1½ pounds cooked and cleaned shrimp

1. In a medium-size skillet over medium heat, melt ¼ cup of the butter. Sauté the banana chunks until lightly browned. Remove bananas from skillet and set aside.
2. In a large skillet melt remaining ¼ cup butter. Add chopped onion and cook until tender. Stir in flour, then add applesauce and boullion, stirring well with a fork until thick and smooth. Stir in curry, ginger, cayenne, and lemon juice.
3. Add shrimp and reserved bananas and continue cooking just long enough to heat shrimp and bananas. Do not overcook. Serve over cooked rice.

Yield: 6 servings

West Indian Stuffed Chicken

This is an old Haitian recipe traditionally made with guinea fowl, but it tastes just as delicious made with Cornish hens, chicken or any small game bird.

Stuffing

3	cups unseasoned dry bread cubes for stuffing
4	large bananas, coarsely mashed
	zest and juice of 1 lime
1	tablespoon dark rum
1	dash Tabasco sauce
¼	teaspoon cinnamon
¼	teaspoon nutmeg
⅛	teaspoon cloves
1	teaspoon salt
1	4-pound roasting chicken

Basting Mixture

¼	cup lime juice
¼	cup dark rum
1 to 2	dashes Tabasco sauce
2	tablespoons vegetable oil

1. Preheat oven to 350° F.
2. In a large mixing bowl, combine all stuffing ingredients except chicken.
3. Rinse and dry the cavity of chicken. Rub with remains of the lime.
4. Stuff chicken loosely and place in a roasting pan. Bake for 1½ hours until done, basting periodically with basting mixture.

Yield: 4 servings

Grouper Key West

This dish combines our local bounty to delicious advantage. If grouper isn't available, it is possible to use another white fish such as sole, cod, or flounder. A good rule of thumb for cooking fish is to allow 10 minutes per inch of thickness measured at the thickest part.

> 2 pounds grouper fillets
> 2 ripe bananas, cut in chunks
> ½ cup mayonnaise
>   zest and juice of ½ lime
> ⅛ teaspoon ground ginger
> ¼ teaspoon white pepper
> ½ cup pecan halves, lightly toasted

1. Preheat oven to 400° F. Grease a shallow baking pan large enough to comfortably hold the fillets. Rinse and dry fillets and place in pan.
2. Place bananas, mayonnaise, lime juice, ginger, and pepper in container of an electric blender. Blend until smooth. Stir in lime zest.
3. Spread banana mixture over fillets and bake for 10 to 12 minutes (see above). Garnish with pecans and thin slices of lime.

Yield: 4 servings

Cakes & Cookies

Banana-Ginger Cream Jelly Roll

Rather than the traditional birthday cake, I made this jelly roll for a friend's birthday party. It offered a truly different taste treat and was a great success enjoyed by all. I must admit it's sinfully rich, but for a special occasion, justifiably permitted.

        7   eggs, separated
        6   tablespoons sugar
        ¾   cup finely ground walnuts
        1   teaspoon ground ginger
            confectioners' sugar
        2   bananas, cut into halves lengthwise and then
                thinly sliced
        ¼   cup orange juice
        ½   cup ginger conserve
     1½   cups heavy cream, whipped

1. Preheat oven to 350° F.
2. Grease a 10 x 15 inch jelly roll pan. Line pan with waxed paper and grease the waxed paper with butter or oil.
3. In a large mixing bowl and using an electric mixer, beat egg whites until stiff.
4. In another large mixing bowl and using an electric mixer, beat yolks long and hard, gradually adding sugar, until mixture is pale and thick.
5. Using a rubber spatula, gently fold beaten egg whites into beaten yolks, incorporating thoroughly but not over blending. Gently fold in ground walnuts and ground ginger.
6. Pour this mixture into prepared pan and, using a spatula, spread evenly to all sides and corners.
7. Bake for 30 to 35 minutes or until lightly browned and cake rebounds to touch.

8. Remove cake from oven. Cover it with a lightweight damp cloth and place on a rack to cool.

9. When cake is cool, remove cloth and sprinkle top with confectioners' sugar. Invert cake onto a large piece of waxed paper. Carefully peel

10. off waxed paper.
Place sliced bananas in a small bowl and sprinkle with orange juice,

11. stirring gently to coat all pieces. Pour off any excess.
Stir ginger conserve into whipped cream until well mixed. Add

12. bananas.
Spread mixture evenly over cake. With your fingers under the waxed paper, roll up the cake like a jelly roll. Sprinkle the top with additional confectioners' sugar.

Yield: 8 to10 servings

The banana plant is a giant herb, the largest plant on earth without a woody stem, and is in the same family as lilies and orchids. There are more than 450 varieties.

Stem: *a fully fruited shoot*
Finger: *a single banana*
Cluster: *has 3 or more fingers*
Hand: *has 8 or more fingers*

Banana-Mocha Cake

Banana, coffee, and cocoa all come from the same region, so it's not surprising they make a delicious combination.

2	very ripe bananas
1	teaspoon instant powdered coffee
1¼	cup all-purpose flour
⅔	cup sugar
¼	cup cornstarch
3	tablespoons powdered cocoa
1	teaspoon baking soda
½	teaspoon salt
1	egg, lightly beaten
⅓	cup vegetable oil
1	tablespoon vinegar
1	teaspoon vanilla

1. Preheat oven to 350° F. Grease and flour a 9 x 9 x 2 inch pan.
2. Slice bananas into a blender. Cover and blend on high speed until smooth. Add coffee and blend again to mix.
3. In a large mixing bowl combine flour, sugar, cornstarch, cocoa, baking soda, and salt. Stir with a fork to mix.
4. Make a well in center of dry ingredients. Add banana mixture, egg, oil, vinegar, and vanilla. Stir with a fork to blend well.
5. Pour into prepared pan and bake for 30 minutes or until wooden tooothpick inserted in center of cake comes out clean. Remove pan from oven and allow to cool on a wire rack for 10 minutes before reinverting the cake out of the pan onto the rack to finish cooling. When cool add frosting.

Creamy Mocha Frosting

3 tablespoons softened butter
1½ cups sifted confectioners' sugar
2 tablespoons powdered cocoa
1 teaspoon instant powdered coffee
½ teaspoon vanilla

In a small mixing bowl using an electric mixer, cream butter. Gradually add remaining ingredients and mix until smooth.

Yield: 9-inch square cake

The "banana crossroads of the world," is located at the border of the twin cities of Fulton, Kentucky, and South Fulton, Tennessee. It celebrates the world's only International Banana Festival. This seemingly unlikely location was, in the early days of the banana trade, a central checkpoint and distribution center for 70 percent of all bananas brought to the U.S.

Mañana Icebox Cake

Make today. Serve tomorrow. This is an old-fashioned icebox cake, but a modern refrigerator will also work.

    8   cups cornflakes
    ½   cup sugar
    ½   cup butter, melted
    4   large ripe bananas, coarsely mashed
    1   cup applesauce
    2   tablespoons dark rum
    ½   teaspoon cinnamon
    ¼   teaspoon nutmeg
    1   cup heavy cream, whipped (not used until cake
        is ready to serve)

1. In a large mixing bowl, crush cornflakes into fine crumbs and combine with sugar.
2. In a large frying pan over medium heat, melt butter.
3. Add crumb mixture to frying pan, stirring over low heat until sugar is dissolved. Remove pan from heat and set aside.
4. In a medium-size mixing bowl mash bananas and stir in applesauce, rum, cinnamon, and nutmeg.
5. Butter a 4-inch deep 1½-quart baking dish.
6. Alternate layers of crumb mixture with banana mixture, beginning and ending with crumbs, gently pressing down each layer.
7. Cover and chill for 24 hours.
8. Serve with whipped cream.

Yield: 6 to 8 servings

Banana–Rum Raisin Bundt Cake

This is a wonderful blend of spices, fruit, and rum for a sensational finale to a company dinner.

½ cup butter	½ teaspoon salt
1½ cups firmly packed light brown sugar	2 teaspoons allspice
	¼ cup milk
3 eggs	¾ cup dark rum
3 medium bananas, coarsely mashed	1 cup golden raisins, tossed in flour to coat
2½ cups all-purpose flour	⅓ cup apricot jam
1 teaspoon baking soda	2 tablespoons dark rum
1 teaspoon baking powder	

1. Preheat oven to 350° F. Grease and flour a 10-inch bundt pan.
2. In a large mixing bowl and using an electric mixer cream butter and sugar. Add eggs one at a time, beating until fluffy. Add bananas and continue to mix until smooth.
3. In a large bowl mix dry ingredients together and beat into banana mixture alternately with milk and rum until well blended.
4. Gently stir in raisins.
5. Pour into prepared cake pan and bake for about 50 minutes or until a wooden toothpick inserted in center of cake comes out clean.
6. Remove pan from oven and allow to cool in the pan for 10 minutes before inverting onto a wire rack to complete the cooling.
7. To glaze: In a small saucepan gently heat apricot jam until melted. Add rum, stirring to mix well. While hot, spoon glaze over cooled cake.

Yield: 12 servings

Banana Upside-Down Cake

¼ cup butter
1 cup firmly packed light brown sugar
1 cup pecan halves
2 ripe bananas, cut lengthwise and crosswise

Cake Batter

¼ cup butter
1 cup sugar
1 egg
   zest of an orange
1½ cups all-purpose flour
¼ teaspoon baking soda
½ teaspoon cream of tartar
1 tablespoon ground ginger
½ cup milk

1. Preheat oven to 350° F.
2. In a 10-inch cast iron skillet melt butter over medium heat and stir in brown sugar; mix well.
3. Remove skillet from heat. Sprinkle on pecans evenly. On top of this, arrange bananas, pinwheel-fashion, with tip ends toward the center and flat sides down. Set aside.
4. Cake batter: In a medium-size mixing bowl and using an electric beater, cream butter and sugar. Add egg and orange zest and beat until fluffy.
5. Combine dry ingredients and beat in with milk until well blended.
6. Pour batter over brown sugar and bananas in iron skillet.
7. Place skillet in oven and bake approximately 25 minutes or until a toothpick inserted in center of cake comes out clean.
8. Remove skillet from oven and immediately turn out cake, upside down, onto a large, heat-proof platter. Cool before serving.

Yield: 8 servings

Josephine's Banana Shortcake

A French twist to an American favorite, this dish is named for Josephine Baker, who immortalized bananas in Paris.

3   large ripe bananas, sliced in rounds
⅓   cup orange juice
2   tablespoons Grand Marnier liqueur

Shortcake

2   cups all-purpose flour
4   tablespoons baking powder
2   tablespoons sugar
½   cup butter
⅔   cup milk
   zest of an orange
1   cup heavy cream, whipped
1   cup orange marmalade

1. Preheat oven to 450° F and grease a cookie sheet.
2. In a medium-size mixing bowl toss bananas carefully with orange juice and Grand Marnier. Refrigerate.
3. In a large mixing bowl mix together dry ingredients and with a pastry blender, cut in butter. Stir in milk and orange zest.
4. On a floured surface, knead dough to mix and then divide into 8 parts. With floured hands, roll each part into a ball and pat down to flatten slightly. Bake on buttered cookie sheet for about 12 minutes or until just golden. Remove cookie sheet from oven and cool shortcake on a wire rack.
5. Slit each shortcake with a fork, spread with a spoonful of marmalade and top with banana mixture. Add whipped cream and more bananas.

Yield: 8 servings

Banana Ginger Wafers

Sort of like a vanilla wafer gone to heaven!

½ cup butter
½ cup sugar
½ cup firmly packed light brown sugar
1 egg
1 large ripe banana, coarsely mashed
2 tablespoons freshly grated ginger root
1 cup all-purpose flour
1 teaspoon baking powder

1. Preheat oven to 350° F. Grease a large cookie sheet.
2. In a large mixing bowl and using an electric mixer, cream butter. Beat in sugars until thoroughly blended. Add egg and beat well to mix.
3. Beat in mashed banana and grated ginger. Add flour and baking powder, half at a time, beating well between additions.
4. Drop by heaping teaspoons 2½ inches apart onto greased cookie sheet and bake for about 12 minutes or until edges are browned and crisp.
5. Remove cookie sheet from oven and remove wafers to wire racks to cool.

Yield: About 4 dozen cookies

Chocolate Chip Tropicals

If you had to imagine a chocolate chip cookie invented in Tahiti, this would be it!

    ½ cup butter
    ½ cup sugar
    ½ cup, firmly packed light brown sugar
    1 egg
    2 ripe bananas, coarsely mashed
    1 cup all-purpose flour
    ½ teaspoon baking powder
    1 cup shredded or flaked sweetened coconut
    1 6-ounce package semi-sweet chocolate chips

1. Preheat oven to 375° F. Grease a large cookie sheet.
2. In a large mixing bowl and using an electric mixer, cream butter. Beat in sugars until thoroughly blended. Add egg and beat well to mix. Add mashed bananas and beat again.
3. Add flour and baking powder, half at a time, beating well between additions.
4. Stir in coconut and chocolate chips with a mixing spoon.
5. Drop by tablespoon 2 inches apart onto greased cookie sheet and bake about 12 minutes or until golden. Remove cookie sheet from oven and remove cookies to wire racks to cool.

Yield: About 2½ dozen cookies

Banana-Oatmeal Cookies

These cookies are so tasty and healthy (and rich in fiber!) you'll have to hide the cookie jar.

¾	cup butter
½	cup sugar
½	cup firmly packed brown sugar
1	egg
1½	cups all-purpose flour
1	teaspoon salt
½	teaspoon nutmeg
½	teaspoon cinnamon
½	teaspoon ground cloves
1	teaspoon baking powder
1	cup rolled oats
2	large ripe bananas, coarsely mashed
1	teaspoon vanilla

1. Preheat oven to 375° F. Grease a large cookie sheet.
2. In a large mixing bowl and using an electric mixer, cream butter. Beat in sugars until well mixed. Add egg and continue mixing until fluffy.
3. Add flour half at a time, along with salt, spices, and baking powder, beating until thoroughly mixed.
4. Mix in oats, mashed bananas, and vanilla until well blended.
5. Drop by tablespoon 2 inches apart onto greased cookie sheet and bake 12 to 15 minutes or until golden Remove cookie sheet from oven and remove cookies to wire racks to cool.

Yield: About 3 dozen cookies

Peanut Butter & Banana Cookies

These cookies are lighter in texture than traditional peanut butter cookies.

¼	cup butter
½	cup sugar
½	cup firmly packed brown sugar
1	egg
1	teaspoon vanilla
½	cup chunky peanut butter
1½	cups all-purpose flour
1	teaspoon baking soda
2	medium-size ripe bananas, coarsely mashed

1. Preheat over to 375° F. Grease a large cookie sheet.
2. In a large mixing bowl and using an electric mixer, cream butter. Add sugars and beat well. Add egg and vanilla and beat again. Add peanut butter and beat just enough to mix in.
3. Add flour and baking soda, half at a time, beating between additions to mix well. Stir in coarsely mashed bananas by hand, just to combine.
4. Drop by tablespoon 2 inches apart onto a greased cookie sheet. Bake about 12 minutes or until lightly browned. Remove cookie sheet from oven and remove cookies to wire rack to cool.

Yield: About 2½ dozen cookies

Puddings & Pies

Banana-Orange Chocolate Pie

Orange and chocolate are a classic combination. The surprise here is the hidden bananas.

Crust
1¼ cups chocolate wafer cookie crumbs (about 24 cookies)
¼ cup melted butter
¼ cup sugar

1. Preheat oven to 325° F. and grease a 9-inch pie pan.
2. In a medium-size mixing bowl combine all ingredients and mix well using a fork.
3. Press cookie-crumb mixture firmly into bottom and sides of pie pan and bake for 10 minutes. Remove pan from oven and cool thoroughly before filling.

Filling
1 6-ounce can frozen orange juice concentrate, thawed
1 14-ounce can sweetened condensed milk
zest of one orange
1 8-ounce package cream cheese at room temperature
2 ripe bananas, sliced

1. Place orange juice, sweetened condensed milk, orange zest, and cream cheese cut into chunks into container of an electric blender. Cover and blend on medium speed for several seconds. Stop motor and push mixture into blades as needed. Cover and blend on high speed for about 30 seconds or until well mixed and smooth.
2. Line bottom of cooled pie crust with banana slices. Pour in blended mixture. Chill well before serving.

Yield: 9-inch pie

Havana Banana Cream Pie

Castro never had it so good! This pie is as sinfully rich as it is delicious.

Crust

1⅓	cups finely crushed graham cracker crumbs
½	cup coarsely chopped pecans
1	teaspoon cinnamon
¼	cup sugar
6	tablespoons butter, melted

1. Preheat oven to 350° F. and butter a 9-inch pie pan.
2. In a medium-size mixing bowl combine all ingredients, mixing well with a fork.
3. Press mixture into bottom and sides of pie pan.

Filling

1	cup heavy cream
2	eggs
3	tablespoons all-purpose flour
⅓	cup dark rum

1. In a mixing bowl, beat all ingredients together until well blended.
2. Pour into crust and bake approximately 20 minutes or until custard filling is set, or when a knife blade inserted in center of pie comes out clean. Remove pan from oven and allow pie to cool thoroughly before adding topping. Pie can also be refrigerated and served cold.

Topping

2 medium bananas, thinly sliced
¼ cup orange juice

Dip banana slices into orange juice to prevent discoloration. Drain before arranging on top of pie. Starting at outside edge and working toward middle, overlap sliced bananas slightly.

Yield: 9-inch pie

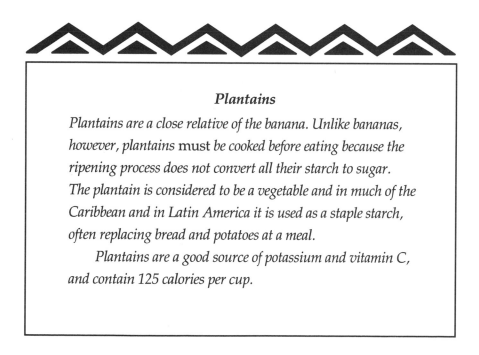

Plantains

Plantains are a close relative of the banana. Unlike bananas, however, plantains must be cooked before eating because the ripening process does not convert all their starch to sugar.

The plantain is considered to be a vegetable and in much of the Caribbean and in Latin America it is used as a staple starch, often replacing bread and potatoes at a meal.

Plantains are a good source of potassium and vitamin C, and contain 125 calories per cup.

Banana Daiquiri Pie

This light pie offers a luscious marriage of tropical flavors!

Crust

2½ cups sweetened, flaked coconut
¼ cup very soft butter

1. Preheat oven to 300° F.
2. Mix coconut and butter together and press into an ungreased 9-inch pie pan.
3. Bake for 25 minutes or until golden brown. Remove pan from oven and cool thoroughly before filling.

Filling

1 envelope unflavored gelatin
⅓ cup light rum
¼ cup boiling water
3 tablespoons lime juice
½ cup sugar
2 very ripe bananas, cut in chunks
1 cup whipping cream or one 8-ounce container Cool Whip
½ teaspoon lime zest

1. Pour rum into electric blender container. Sprinkle gelatin over rum and allow to dissolve for 5 minutes.
2. Add boiling water, cover, and blend on high speed for 30 seconds. Add lime juice and sugar. Continue blending for another 30 seconds. Add bananas and blend until smooth.
3. Empty banana mixture into a large mixing bowl and refrigerate for about 20 minutes or until syrupy.
4. In a small, chilled mixing bowl beat whipping cream until fluffy and thick, but not too stiff.

5. Fold whipped cream and lime zest into chilled banana mixture.
6. Pour into prepared crust and chill again before serving.

Yield: 9-inch pie

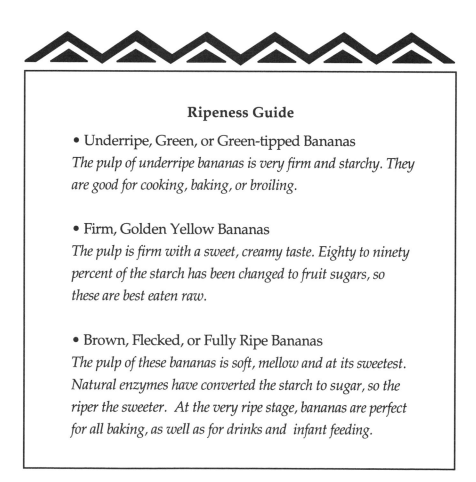

Ripeness Guide

• Underripe, Green, or Green-tipped Bananas
The pulp of underripe bananas is very firm and starchy. They are good for cooking, baking, or broiling.

• Firm, Golden Yellow Bananas
The pulp is firm with a sweet, creamy taste. Eighty to ninety percent of the starch has been changed to fruit sugars, so these are best eaten raw.

• Brown, Flecked, or Fully Ripe Bananas
The pulp of these bananas is soft, mellow and at its sweetest. Natural enzymes have converted the starch to sugar, so the riper the sweeter. At the very ripe stage, bananas are perfect for all baking, as well as for drinks and infant feeding.

Clafoutis aux Bananas Tropicale

A French classic with a tropical twist, this recipe uses bananas and coconut rather than apples or cherries. You might find this dessert in Martinique.

3 large bananas, cut into halves lengthwise	1 cup half & half
½ cup shredded sweet coconut	¼ cup dark rum
	½ cup sugar
	½ cup all-purpose flour
3 eggs	⅛ teaspoon nutmeg

1. Preheat oven to 375° F. Grease a 10-inch pie pan.
2. Arrange banana halves in prepared pan and sprinkle with coconut.
3. Place remaining ingredients in the container of an electric blender. Cover and blend on high speed for one minute, stopping after 30 seconds to scrape down the sides.
4. Pour batter over bananas and bake for about 45 minutes or until golden. Remove pan from oven. Serve warm or cooled.

Yield: 6 servings

International Banana Festival's One Ton Banana Pudding

3,000 bananas, sliced

250 pounds vanilla wafers

950 pounds banana pudding and pie filling

Fill a 5-foot-wide and 3-foot-deep bowl with sliced bananas. Spread alternating layers of vanilla wafers and pudding. **Serves 10,000!**

Top Banana Bread Pudding

Take grandma's favorite bread pudding and add bananas for this updated, delicious version.

4	teaspoons softened butter
8	slices cinnamon raisin bread
2	large bananas, cut into half-inch slices
3	eggs
2	cups half & half
½	cup sugar
2	tablespoons sherry
¼	teaspoon nutmeg
1	teaspoon vanilla

1. Grease a 2½-quart, deep baking dish such as a soufflé dish. Butter slices of bread.
2. With buttered sides down, make layers of bread and bananas, starting and ending with a bread layer, in the prepared baking dish.
3. Place remaining ingredients in container of an electric blender. Cover and blend on high speed for 30 seconds. Slowly pour mixture over bread and allow to stand for 30 minutes. Dot with butter.
4. Preheat oven to 350° F.
5. Cover loosely with foil and bake for 30 minutes. Remove foil and continue baking another 15 minutes or until golden and puffed. Remove dish from oven. Serve warm or chilled.

Yield: 8 servings

Sherry Custard Bananas

A custard fit for a queen!

4 large bananas, cut into ½-inch slices
juice of 2 lemons
2 large eggs, separated
¼ cup sherry
¼ cup sugar
zest of 1 lemon
¼ teaspoon nutmeg
½ cup heavy cream, whipped

1. Preheat oven to 325° F. Grease an 8 x 8 inch baking dish.
2. Arrange banana slices in prepared pan and sprinkle with lemon juice. Set aside.
3. Carefully separate eggs, putting yolks into a large mixing bowl and whites into a medium mixing bowl.
4. Using an electric mixer beat the whites until stiff but not dry.
5. In a separate bowl, beat yolks and sherry thoroughly. Add sugar, lemon zest, and nutmeg, and continue beating until frothy and pale in color.
6. Using a rubber spatula gently fold egg whites into yolk mixture, then fold in whipped cream.
7. Pile mixture onto bananas and bake for 30 minutes. Remove dish from oven. Serve warm.

Yield: 6 servings

Banana Maple Charlotte

Bananas and ladyfingers combine to make an elegant, old-fashioned dessert.

¼ cup cold water
1 packet unflavored gelatin
¾ cup very hot maple syrup
2 large ripe bananas, cut in chunks
1 package ladyfingers
1 cup heavy cream, whipped
½ cup broken walnut pieces

1. Place cold water in container of an electric blender. Sprinkle on gelatin and let dissolve for 5 minutes.
2. Add hot maple syrup. Cover and blend at high speed for 30 seconds. Add banana chunks and blend again until smooth.
3. Pour mixture into a large mixing bowl and chill until syrupy, approximately 45 minutes.
4. Line a 1-quart mold or loaf pan with split ladyfingers, rounded sides to the outside/bottom.
5. Fold whipped cream and walnuts into the banana mixture and pour into lined mold. Chill overnight. Unmold onto a serving plate.

Yield: 6 to 8 servings

Other Desserts,
Frozen to Flambéed

No-Cream Banana "Ice Cream"

You can not only put bananas in the refrigerator to hold them at a desired degree of ripeness for 1 to 2 days, but you can also freeze them for several months.

Bananas can be defrosted or used partially frozen in any recipe calling for mashed bananas.

To freeze: Peel and dip whole bananas in lemon or lime juice to prevent discoloration. Wrap individual bananas snugly in plastic wrap so they are covered completely, thus eliminating any air space.

   With a stash of bananas in your freezer it is possible to whip up a great variety of instant banana "ice creams." The combinations are myriad. Use your imagination. The good news is that many of these recipes are virtually fat- and cholesterol-free and very taste-satisfying.

Banana-Strawberry "Ice Cream"

4 large frozen bananas, cut into chunks
1 cup hulled and halved strawberries
2 tablespoons Grand Marnier liqueur (optional)

Using either a blender or food processor with a steel blade, process all ingredients until smooth. This produces a soft "ice cream" to be served at once, or to be stored in your freezer.

Yield: 4 servings

NOTE: This recipe may be varied by substituting any of the following for the strawberries: fresh, canned, or frozen peaches, raspberries, apricots, blueberries, pineapple, or cranberry sauce.

Drain off excess syrup from canned or frozen fruit before using.

Banana Daiquiri "Ice Cream"

So quick! So easy! So good!

4 large ripe frozen* bananas, cut into chunks
2 tablespoons lime juice
4 tablespoons light rum

1. Using either a blender or a food processor with a steel blade, process all ingredients until smooth.
2. This produces a soft "ice cream" to be served at once, or to be stored in your freezer.

Yield: 4 servings

*See page 81.

Bananas have no cholesterol and are low in sodium.
Bananas contain tryptophan, an essential amino acid that reduces stress.
Bananas are a significant source of 3 kinds of fiber: crude, nondigestible, and pectin.

Banana-Colada "Ice Cream"

This is a truly tropical dessert—good enough to eat, rather than drink.

1 8¼-ounce can crushed pineapple
¼ cup light rum
3 large frozen bananas, cut into 1-inch chunks
½ cup sweetened flaked coconut

1. Set freezer at coldest setting.
2. In several batches place pineapple, rum, and banana chunks into container of an electric blender. Cover and blend on high between additions, scraping down when necessary until all amounts are blended.
3. Add coconut. Cover and blend on high speed until smooth.
4. Pour into a metal 8 x 8 x 1½ inch pan and freeze. Stir when freezing begins and again just before mixture becomes firm. To serve, scoop as you would ice cream. For a deluxe version, embellish with banana slices and chocolate sauce.

Yield: 6 servings

Banana Frozen Yogurt

It's so easy to make your own frozen yogurt! Try different flavored Jell-O for variety.

1	package strawberry Jell-O
½	cup orange juice, heated
3	large ripe bananas, cut in chunks
3	cups plain yogurt

1. Set freezer at coldest setting.
2. Place Jell-O and hot orange juice in container of an electric blender. Cover and blend at high speed for 30 seconds.
3. Add banana chunks a few at a time and then yogurt. Continue blending until smooth.
4. Pour into a metal loaf pan and freeze. Stir when freezing begins and again just before the mixture becomes firm. To serve, scoop as you would ice cream. Garnish with your choice of fresh or frozen fruit if desired.

Yield: One quart

Frozen Banana Fandango

This is our tropical version of the icebox cake.

 ½ cup orange juice
 16 marshmallows
 1 cup vanilla yogurt
 4 large ripe bananas, cut in chunks
 30 graham cracker squares

1. Place orange juice and marshmallows in container of an electric blender. Cover and blend on high speed, scraping down when necessary. Add yogurt and banana chunks. Cover and blend again on high speed about 30 seconds until smooth.
2. Arrange 15 graham crackers in a 9 x 13 x 2 inch pan. Pour half banana mixture over graham crackers. Repeat with remaining banana mixture and then with remaining graham crackers.
3. Freeze until firm. Cut into squares.

Yield: 15 servings

Banana Fool

Bananas and strawberries combine to make this old-fashioned dessert that was transported from England to its West Indian colonies.

  1   package lemon Jell-O
  1   cup boiling water
  4   large, ripe bananas, cut in chunks
      zest of a lemon
  ½   cup heavy cream
  1   pint strawberries, sliced (reserve 6 of the
          prettiest ones for garnish)

1. Place Jell-O and boiling water in container of an electric blender. Cover and blend at high speed for 30 seconds. Add banana chunks, lemon zest, and cream. Cover and blend until smooth. Pour into a medium bowl.
2. Chill until thickened (approximately 2 hours), stirring occasionally.
3. Spoon into sherbet glasses, making alternate layers with strawberry slices. Garnish with a whole strawberry. Chill until firm, at least 1 hour.

Yield: 6 servings

A Trifle Tropical

This is a traditional English dish, adapted for the tropics.

¼ cup confectioners' sugar
1 cup heavy cream, whipped
3 large, ripe bananas, coarsely mashed
½ cup sherry
1 package ladyfingers
1 cup pineapple preserves
1 cup sweetened flaked coconut

1. In a large mixing bowl combine whipped cream with sugar.
2. Add bananas to sherry. Beat well to combine.
3. Split ladyfingers and spread with pineapple preserves. Divide equally and arrange among 6 glass dessert dishes.
4. Spoon on a layer of banana cream mixture and then the equally divided coconut. Add remaining banana cream mixture. Chill well before serving, at least 2 hours or overnight.

Yield: 6 servings

Banana Toffee Crunch

This dessert is outrageously naughty, but you deserve a little fling every now and then.

6	medium bananas, diced
½	cup sugar
½	teaspoon nutmeg
1	cup firmly packed brown sugar
1¼	cups all-purpose flour
1	cup rolled oats
½	cup chopped pecans
¾	cup butter

1. Preheat oven to 375° F. and grease an 8 x 8 x 1½ inch pan.
2. In a medium-size mixing bowl, mix diced banana, ½ cup sugar, and nutmeg. Pour into prepared baking pan.
3. In a large mixing bowl combine brown sugar, flour, oats, and pecans, stirring well to mix. Cut in butter with a pastry blender until crumbly. Sprinkle over bananas.
4. Bake for about 30 minutes or until crisp and browned.
5. Serve warm with cream or ice cream.

Yield: 6 servings

Bananas Poached in Red Wine

This sophisticated combination of flavors is the perfect grace note to an elegant meal.

    1   cup red wine
    1   cup orange juice
        zest of an orange
    1   stick cinnamon
    1   teaspoon ground allspice
    3   tablespoons confectioners' sugar
    4   small, whole bananas
    1   pint orange sherbet

1. In a large, heavy saucepan combine all ingredients except the bananas and sherbet.
2. Bring to a boil over medium heat, then reduce heat and simmer for 5 minutes. Add bananas and simmer gently for another 5 minutes or until bananas are soft.
3. Carefully remove bananas and boil liquid rapidly to reduce by half. Return bananas to liquid and chill for at least 1 hour.
4. Serve with orange sherbet.

Yield: 4 servings

Jamaican Banana Crisp

Gingersnaps provide the crunch to this taste of tropical splendor

6	medium bananas, diced
⅓	cup sugar
½	teaspoon ground ginger
½	cup pineapple juice
16	gingersnaps, crushed into crumbs
¼	cup butter, softened
½	cup shredded or flaked sweet coconut
2	tablespoons light brown sugar

1. Preheat oven to 350° F. Grease a shallow 2-quart baking dish.
2. In a medium mixing bowl toss bananas with ⅓ cup sugar and ground ginger to coat. Add pineapple juice. Pour into prepared pan.
3. In a medium mixing bowl combine gingersnap crumbs, butter, coconut, and light brown sugar stirring well with a fork to blend. Sprinkle over bananas.
4. Bake for 25 to 30 minutes until crisp and bubbly. Remove dish from oven and serve warm.

Yield: 6 servings.

Rum Hard Sauce

Jamaican Banana Crisp is especially good when served with Rum Hard Sauce.

To make sauce: In a small mixing bowl and using an electric mixer, cream ¼ cup butter. Then gradually add 1 cup confectioners' sugar and 1 teaspoon dark rum, and beat well until combined.

Kahlúa-Carmelized Bananas

Indescribably delicious!

⅓	cup butter
⅓	cup firmly packed light brown sugar
3	bananas, cut into ½-inch slices
¼	cup Kahlúa liqueur
½	cup pecan halves, toasted
1	pint coffee or chocolate ice cream

1. In a large, heavy skillet or chafing dish, heat butter and brown sugar together over medium heat until sugar is melted. Continue to cook over a low heat for about 5 minutes or until sugar begins to caramelize.
2. Add bananas. Stir gently to glaze and heat through. Add Kahlúa and pecans.
3. Serve hot over ice cream.

Yield: 4 servings

A Connecticut used-car dealer advertised and sold a car for 600 bananas.

Bananas Foster

This dish was created at Brennan's Restaurant in New Orleans in the 1950s and named for Richard Foster, a regular customer at this legendary restaurant.

2 tablespoons lemon juice
4 small bananas, cut in half lengthwise and crosswise
½ cup firmly packed light brown sugar
4 tablespoons butter
½ teaspoon cinnamon
2 tablespoons dark rum
2 tablespoons brandy or cognac
2 tablespoons banana liqueur
1 pint vanilla or coffee ice cream

1. Drizzle lemon juice over bananas.
2. In a large heavy skillet or chafing dish, heat brown sugar and butter together over medium heat until sugar is melted and bubbly.
3. Add bananas and sprinkle on cinnamon. Cook slowly a minute or two until bananas are well heated and glazed.
4. Add rum, brandy, and banana liqueur and heat for another minute. Stand back a bit and ignite. Using a long-handled spoon, ladle liquid over bananas until flame dies out.
5. Serve warm over ice cream.

Yield: 4 servings

Kids' Treats

Jungle S'Mores

Bananas and peanut butter give this old favorite a new twist. These are so good the grown-ups will want them, too.

8	graham crackers
¼	cup peanut butter
4	small ripe bananas, cut in half lengthwise and crosswise
4	Hershey chocolate bars (1.2-ounce size), broken in halves
8	marshmallows*

1. Spread graham crackers with peanut butter and break into halves.
2. Place 2 pieces of banana on each graham cracker half. Top each with half a chocolate bar.
3. Over a campfire, barbecue, or over a stovetop gas flame, toast the marshmallows until golden.
4. Place one marshmallow on top of the chocolate and top with half a graham cracker, pressing together lightly.

Yield: 8 treats

*If a campfire isn't handy, use 1 tablespoon marshmallow Fluff per S'More instead.

Banana Bonbons

No bake and so easy, even a kid can make them!

  2   cups finely crushed vanilla wafers
½   cup finely chopped peanuts
¼   cup sweetened, flaked coconut
  1   ripe banana, mashed
¼   cup confectioners' sugar
  1   teaspoon powdered cocoa

1. In a medium-size mixing bowl, combine vanilla wafer crumbs, peanuts, coconut, and banana. Mix well with a fork to blend. Shape into 1-inch balls.
2. Into a shallow bowl, sift together confectioners' sugar and cocoa powder. Roll balls in this mixture to coat evenly.
3. Store in an airtight container and hide!

Yield: approximately 3 dozen bonbons

Banana Facial Mask

Mash one very ripe banana together with 2 teaspoons olive oil. Apply to face and wait 15 minutes before rinsing off thoroughly with clear water. Pat dry.

Birthday Candle Salad

You might remember this recipe from your childhood. Kids still love it.

> 2 large bananas, cut in halves crosswise
> 1 8-ounce can pineapple slices drained
> (reserve juice)
> 2 maraschino cherries, halved
> 16 strawberries
> iceberg lettuce

1. On a plate, place a pineapple slice on a bed of lettuce leaves.
2. Place half a banana, cut side down, in center of pineapple slice to form a "candle." Add a cherry half for the "flame." Arrange 4 strawberries in lettuce surrounding pineapple.

Yield: 4 servings

Banana Nut Mayonnaise

To complement Birthday Candle Salad make Banana Nut Mayonnaise by mixing ½ cup mayonnaise, 2 tablespoons reserved pineapple juice, 1 tablespoon peanut butter, and ½ ripe mashed banana, until well blended. The Grenadine Dressing on page 45 is another good choice to serve with the salad.

Banana Boat Salad

Sailor's delight! A banana boat sailing on a lettuce sea that's good enough to eat. The Grenadine Dressing on page 45 is a good and pretty addition to this salad.

    4   large bananas, halved lengthwise (Then cut
        one of the halves in half lengthwise and
        crosswise.)
    1   16-ounce can fruit cocktail, drained
    2   cups cottage cheese
        iceberg lettuce

For each of four salads:
1. On an 11-inch plate, arrange a "sea" of lettuce leaves on lower portion of plate.
2. Place a banana half on surface of "sea," creating the hull of a sailboat.Using remaining full-length piece of banana, create the "mast." Using 2 shorter pieces of banana, outline the "sail."
3. Fill in the "sail" with some of drained fruit cocktail.
4. Spoon ½ cup cottage cheese onto lettuce.

Yield: 4 servings

Banana Fruit Pops

This is a wholesome and economical treat, the perfect combination.

    4   small, ripe bananas, in chunks
    1¼ cups fruit juice, such as orange, pineapple,
        cranberry, apricot, or grape
    8   wooden popsicle sticks

1. Place bananas and fruit juice of your choice in container of an electric blender. Cover and blend on high speed until smooth.
2. Divide mixture evenly among six to eight 4-ounce paper or plastic cups.
3. Place in freezer and when partially frozen place a popsicle stick in center of each cup. Freeze until firm. Let stand at room temperature for 5 minutes before removing cups and serving.

Yield: 6 to 8 popsicles

The per capita consumption of bananas in the U.S. is just over 26 pounds per year. That comes to about 11.5 billion bananas, or 78 per every man, woman, and child. In Eastern Europe it is ½ pound per person per year.

Banana Blitz Cupcakes

No frosting, but look for the surprise!

2	3-ounce packages cream cheese at room temperature
1⅓	cups sugar divided
1	egg
1	6-ounce package semi-sweet chocolate chips
3	large ripe bananas, 2 cut in chunks, 1 sliced
⅓	cup vegetable oil
1	teaspoon vanilla
1½	cups all-purpose flour
¼	cup powdered cocoa
1	teaspoon baking soda
½	teaspoon salt

1. Preheat oven to 350° F. and line muffin tins with cupcake papers.
2. In a small mixing bowl and using an electric mixer, beat cream cheese and ⅓ cup sugar until light and fluffy. Add egg and beat again until well mixed. Stir in chocolate chips by hand. Set aside.
3. Place 2 bananas, broken into chunks, vegetable oil, and vanilla into an electric blender. Cover and blend until banana is puréed.
4. In a large mixing bowl, combine flour, 1 cup sugar, cocoa, baking soda, and salt. Add puréed mixture to dry ingredients, stirring with a fork until well mixed.
5. Divide batter evenly into 18 lined muffin tins and top each with 2 slices of banana. Top this with equal amounts of the cream cheese-chocolate chip mixture.
6. Bake for 30 minutes. Remove muffin tins from oven and cool for 10 minutes before removing cupcakes to a wire rack to cool completely.

Yield: 18 cupcakes

Resources

International Banana Club Headquarters
2524 North El Molino Avenue
Altadena, California 91001   (818) 798-2272
Top Banana: Ken Bannister

Fulton-South Fulton International Banana Festival
Marcy Dement, Executive Secretary
P.O. Box 428
Fulton, Kentucky 42041   (502) 472-2975

Rare Fruit Council International
P.O. Box 561914
Miami, Florida 33256   (305) 238-2809
An affiliate of the Museum of Science in Miami, the Rare Fruit Council is a major international source of information regarding tropical fruits. Membership is open to anyone with an interest in rare fruits and includes a subscription to "Tropical Fruit News." They offer a wide range of activities from field trips and fruit tastings to seed and plant exchanges, as well as the *Tropical Fruit Cookbook.*

The Banana Tree, Incorporated
715 Northampton Street
Easton, Pennsylvania 18042   (215) 253-9589
If you want to grow your own, they offer 23 varieties of banana plants by mail. Catalog available.

W. O. Lessard Nursery
19201 S.W. 248th Street
Homestead, Florida 33031   (305) 247-0397
Offers 51 varieties of banana plants by mail. Catalog available.

Index